# Java

The Ultimate Guide to Learn Java and C++ (Java for beginners, Java for dummies, how to program, C programming, c plus plus, programming for beginners)

# Java

## THE ULTIMATE GUIDE TO LEARN JAVA PROGRAMMING FAST

### PETER HOFFMAN

# Java
## The Ultimate Guide to Learn Java Programming Fast (Java for Beginners, Java for dummies, how to program, java apps, java programming)

PETER HOFFMAN

Copyright © 2016 Peter Hoffman

All rights reserved.

ISBN: 152347081X
ISBN-13: 978-1523470815

# CONTENTS

Introduction ............................................................................................. 7

Chapter 1 – Writing a Java Program ................................................... 10

    **Software Types** ............................................................................... 10

    **Classes, Instances, and Objects** ................................................... 12

    **Project Creation** ............................................................................ 13

    **Main Class Creation** ..................................................................... 13

    **Writing the Package Statement** ................................................... 15

    **Writing the Class Declaration** .................................................... 16

    **Writing the Main Method** ............................................................ 17

    **Writing the Program Output** ...................................................... 18

    **Allowing User Input** .................................................................... 20

Chapter 2 – Variables ............................................................................ 23

    **Overview of Data Types** .............................................................. 25

    **Creation of Variables** ................................................................... 27

    **Data Type and User Input** .......................................................... 29

    **Storing User Input into Another Variable** ................................ 31

Chapter 3 – Operators ........................................................................... 33

    **Summary in a Single Program** .................................................... 34

Conclusion .............................................................................................. 37

I think next books will also be interesting for you:

C++

Python

Javascript

Computer Hacking

Hacking for Dummies

## Amazon Prime and Kindle Lending Library

# Introduction

The company behind the Java programming language is the Sun Microsystems. The language was initially introduced in the year 1991. It was developed by a team of people consisting Mike Sheridan, Ed Frank, Chris Warth, Patrick Naughton, and James Gosling. Due to its object-oriented nature, the Java language can be reused with regards to data focus, classes creation, written instructions, and the creation of both instances and objects.

## The Four Java Platforms

There are four primary platforms in the Java programming language. They are as follows:

- Java Standard Edition (Java SE)
- Java Enterprise Edition (Java EE)
- Java Micro Edition (Java ME)
- JavaFX

Java-based applications that were being used for computer software were developed on the standard edition of Java. Java-based applications for web servers were developed on the enterprise edition while Java-based applications for multimedia platforms were developed on JavaFX. Most JavaFX-based applications are what we usually call flash players. Lastly, Java-based applications for mobile devices were developed on the micro edition Java platform.

## Tools for Beginners

In order to begin your journey to the world of Java programming, you need to choose which IDE to use. IDE, also known as Integrated Development Environment, serves as your workspace in writing Java applications. Imagine it as your notepad in writing HTML codes. However, IDEs were specifically designed for the Java language which means it features all the tools you might need in Java development.

Here are the following IDEs that you can use:

- BlueJ – this is an ideal IDE for beginners. You can easily navigate due to its simplistic interface. It was also originally designed to help beginners in familiarizing the primary concepts of the Java language.

- DrJava – this is a lightweight IDE. Similar to BlueJ, DrJava was also designed for beginners. Although it features a simplistic user interface, DrJava has enough tools that every advanced Java developer can use as well.

- Eclipse – one of the popular IDEs today, the Eclipse is an open source platform with responsive navigation and design. Apparently, it has the cleanest user interface compared to other IDEs in this list.

- NetBeans – this is another popular IDE which is commonly used in Java development nowadays. This is highly recommended for beginners due to its fast and powerful tools that are compatible in all platforms of the Java language.

Each IDE has a unique user interface and ease of navigation. Feel free to use any IDE where you feel most comfortable with. Keep in mind that all of them use the same language although they may appear to be different from one another.

Aside from Java IDEs, another tool you need to begin learning the Java programming language is the JDK. JDK, which stands for the Java Development Kit, serves as your computer's interpreter and compiler for the Java language. Without this, your computer will not be able to translate Java codes into the universal machine language. All versions of JDK are available from the official website of Oracle.

Take note that you need to choose the JDK version that is most appropriate to your computer's operating system to avoid future issues when compiling written Java programs.

## Misinterpretations

Before we proceed, I would like to clarify that the Java language is not related to the JavaScript language. These are two different programming languages whereas Java is used for the development of stand-alone applications. On the other hand, the JavaScript language is an integrated language for web development. It is used to assert functions that are not achievable through HTML language alone.

This book is about the Java programming language and not about the JavaScript language.

# Chapter 1 – Writing a Java Program

Once you are done from choosing your preferred IDE and from downloading the latest version of JDK, you now have the tools to begin programming using the Java language. However, having the tools does not mean that you are already prepared to jump into writing your very first Java application.

It is also important to understand the structure of a Java program – its primary components and their components in the Java application.

**Software Types**

You can create two types of software using the Java programming language – applets and applications.

Applets are smaller pieces of programming codes that were designed to run on web browsers. We can consider the applets as smaller and lighter applications. They are mainly used to provide either navigation enhancements or additional interactivity to the browser. In contrast with stand-alone Java applications, applets do not need any interpreter in order to execute.

Another type of software that you can develop through Java is a console application. Console applications are stand-alone programs that run within a console environment. All IDEs feature an integrated console environment in order for you to test these kinds of programs. As a beginner, it is important for you to learn console operations to have a good grasp on the basics of Java language.

## Classes, Instances, and Objects

The classes, instances, and objects are the three vital elements of a program created using an object-oriented language such as the Java. In order to understand these three, you need to first look at their relationship from one another.

Classes are the highest and they cover everything in Java programming language. Specific elements in a class are called Objects. Once you add the additional specification to an Object, then it becomes an Instance.

In an analogy, for instance, let us say our class here is the animal. Under animal class, we have different kinds of animals like dogs, cats, birds, and so on. These are the objects under animal class. Under the category of dogs, we have different breeds like poodles, golden retrievers, Labradors, and so on. These are the instances of the object dogs.

## Project Creation

The first step in writing your Java program is to start a new empty project in your chosen IDE. You can do this by navigating through the top most toolbar as follows:

*File -> New Project*

After selecting New Project, a new window will appear where you will be able to choose from a list of categories. Under the Categories list, select Java and then select Java Application from the adjacent window panel. Afterward, click on the Next button and you will be asked for the project name and the location where you want to save your project.

Always remember the project name since you will be asked to go back to the same project along the way.

## Main Class Creation

After the creation of a new project, the IDE will automatically create both the package and the main class. Packages, also known as source packages, serves as folders where you can efficiently organize all assets of the Java project such as classes and instances. Take note, however, that packages are simply identifiers for the organization of assets. Consider them as folders in your computer where you organize your files accordingly.

There should be a main class under the package. If the IDE did not automatically create the main class, then you can create it by navigating on the main menu as follows:

*File -> New File -> Java Class*

The main class has its vital role in a Java program. The main method, which is what initially executed when opening a Java program, is contained in the main class. In other programming languages, the main method is known as the entry point.

## Writing the Package Statement

Now that we already have the main class in our Java project, we are now ready to write our first Java programming language line – the package statement. This statement is an optional part of the program and is only introduced to you for learning purposes. This is written using the following syntax:

package [package name];

The IDE will automatically look for the default package if this line is not written. In other words, the package statement might come in handy if you are working on a Java project with multiple packages. Keep in mind that package name must match with the name of the package where the class you need belongs to. Let us say, for instance, that the name of your package is "javatestproject". The syntax should look like this in your IDE:

```
1
2    project javatestproject;
3
```

## Writing the Class Declaration

Next to the package statement is the class declaration. The syntax we need to follow here is this:

[access modifier] class [class name]

There are two kinds of access modifiers that you can use: public and private. The private modifier prevents the other classes from accessing the declared class while the public modifier allows other classes to access the declared class. For the meantime, let us focus and use the public as our access modifier for our first Java application.

Class name, as what the name suggests, is the name you designated to the main class. Take note that the Java programming language is case sensitive so make sure that you name of the main class with correct upper case and lower case characters. Let us say that the name of our main class is "JavaTestProject", then this is what should we have right now:

```
1
2    package javatestproject;
3
4    public class JavaTestProject {
5
6    }
7
```

The open ( { ) and close ( } ) curly braces refer to the opening and closing of a Java code block. This helps the program in determining where a certain block of code begins and where it ends.

## Writing the Main Method

Now that we are done declaring the class, it is now time to declare the main method for our first program. In our main method, we will be using two keywords: void and static.

The void keyword signals the Java Virtual Machine, also known as JVM, that the program successfully ran and finished. The keyword static signals the program that the field refers to the class and neither the objects nor the instances in it. In other words, the program will be able to freely go through the class without creating instances in that class. Due to the nature of the main method to act as the program's entry point, it is essential to declare it as static. Hence, our syntax for the main method declaration should follow this format:

[access modifier] static void

However, this is not a complete declaration for the main method yet. We should also add the parameter "main(String[] args)" to allow the program in executing command-line arguments. And so, our final code for the main method should look like this:

public static void main(String[] args)

If added to our line of codes above, our IDE should look like this:

```
1
2    package javatestproject;
3
4    public class JavaTestProject {
5
6        public static void main(String[] args) {
7
8        }
9
10   }
11
```

## Writing the Program Output

As of now, you already covered the basic parts of our Java program. Unfortunately, we are not yet telling the program what exactly it needs to do.

What we want for the program to execute is to deliver a message. This is what we call as the Program Output or Output Stream. In order for the program to deliver a message, we should use the following syntax:

System.out.print("[message]");

Instead of the parameter "print", you may replace it with "println". This will print the output to a new line once executed. Hence, the alternative format should look like this:

System.out.println("[message]");

Let us now tell the program to print the statement "Java is Awesome!" using the command line above. Our syntax should be like this:

System.out.print("Java is Awesome!");

And so, our IDE should look like this once we insert the output code:

```
1
2   package javatestproject;
3
4   public class JavaTestProject {
5
6       public static void main(String[] args) {
7
8           System.out.print("Java is Awesome!");
9
10      }
```

```
11
12  }
13
```

## Allowing User Input

The Java programming language allows you to let users communicate with the program by sending inputs. We can do this by using the built-in class known as the Scanner.

The Scanner class collects the user's input from the computer's keyboard. It stores the collected data into variables that can also be executed in the program's output stream. In order to do this, we need to add the following lines right after the package statement:

import java.util.Scanner;

This is an example of an import statement which signals the program to import the Scanner class from the built-in Java package known as java.util. This is where we add them in our IDE:

```
1
2   package javatestproject;
3   import java.util.Scanner;
4
```

This, however, does not do anything yet. We need to assign a variable name to the scanner class first in order to call the scanner's collected data into our output stream. We can assign a variable name to the scanner class by using the following statement:

Scanner [variable name] = new Scanner(System.in);

You can assign any name to the variable. It is a common practice to start the name with an uppercase character followed by lowercase characters. It is also a good practice to separate words by using uppercase characters. Let us say that we want to name our scanner variable as "userinput". Instead of writing the entire variable name in lowercases, replace the first letters of the words "user" and "input" to uppercases. Hence, the statement will look like this:

Scanner UserInput = new Scanner(System.in);

We should insert the statement within the blocks of code after the declaration of the main method. Hence, our IDE should be as follows:

```
1
2    package javatestproject;
3    import java.util.Scanner;
4
5    class JavaTestProject {
6
7        public static void main(String[] args) {
8
9            Scanner UserInput = new Scanner(System.in);
10
11       }
12
13   }
14
```

With this set of Java codes, we are commanding the program to assign the user input to the variable named as "UserInput". Let us now tell the program to call the input. As an example, let us try to ask the user's name through our program and let the program itself return the user's name by delivering another output. Our program should be following this schema:

First Output -> Input -> Second Output

In order to create the first output, we will be using what we have learned from writing an output. Let us ask the user's name by asserting the following statement above the scanner assigns variable name statement:

System.out.print("What should we call you?");

Next, let us return the user's answer by calling the collected data into the second output. We can do that by writing the second output below the scanner variable name statement. Our second output should be:

> System.out.println("Hello "+UserInput.nextLine());

With this, we can come up with our final Java program written as follows:

```
1
2   package javatestproject;
3   import java.util.Scanner;
4
5   class JavaTestProject {
6
7       public static void main(String[] args) {
8
9           System.out.print("What should we call you?");
10          Scanner UserInput = new Scanner(System.in);
11          System.out.println("Hello "+UserInput.nextLine());
12
13      }
14
15  }
16
```

# Chapter 2 – Variables

Due to the introduction to the scanner class above, we are now familiar with variables. However, the full potential of variables was still unknown to us.

Variables serve as containers of data. They allow us to call specific data in a more convenient way. We can also create a new set of data using inconsistent values through the use of variables. Most programming languages use variables and developers were able to solve both arithmetic and logical operations through them.

In order to utilize a variable, we need to declare it first and then assign a data type upon its declaration. By doing so, the program will be able to recognize what kind of value was stored inside a variable.

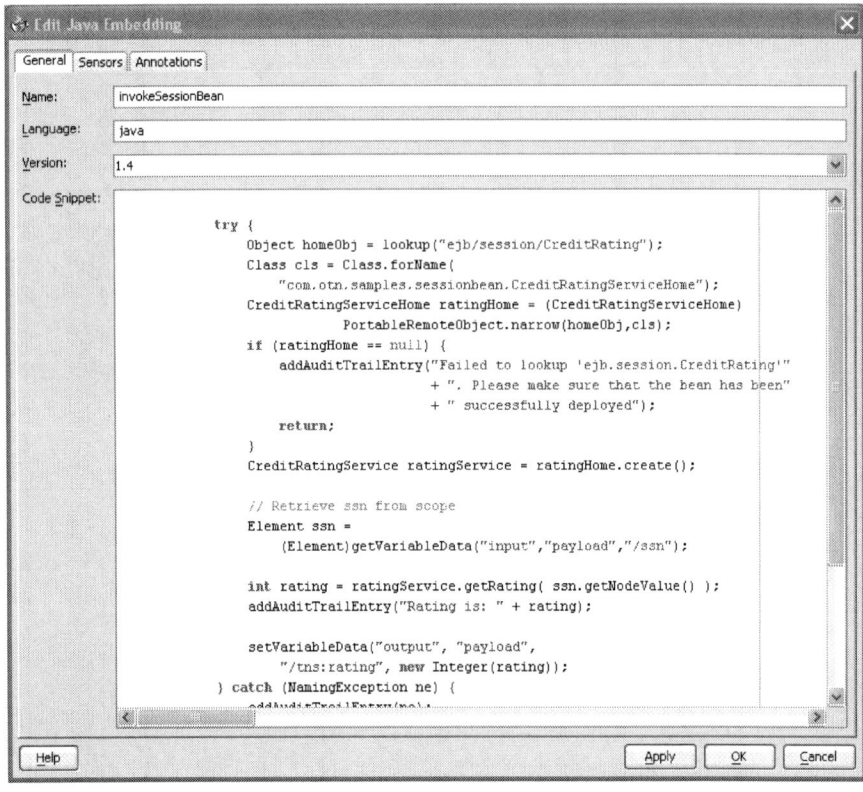

## Overview of Data Types

There are different data types and we can use each type in various operations and functions. Before we can learn how to assign a data type to a variable, it is important to first familiarize ourselves with the data types. Here is a table of the data types we can use for variables in the Java programming language and their brief descriptions:

| Data Type | Description |
|---|---|
| Boolean | This data type has only two possible values: true or false. |
| Byte | This data type has the tiniest range compared to all number-based data types. Its value by default is 0 and it may store an 8-bit 2s integer. |
| Character | This data type only stores a single character. By default, the value will always be an empty space. |
| Double | This data type can store numerical value of 64-bit IEEE 754 and has a value of 0.0d by default. These are numerical values that have decimal points. |
| Float | This data type stores 32-bit IEEE 754 which is less specific than the Double data type. By default, data is set to 0.0f. |
|  | This data type is the simplest way to handle |

| | |
|---|---|
| Integer | numerical values. When not specified, it will automatically translate the value into 0. It does not support decimal points which make it inefficient for precise values. |
| Long | This data type allows you to store numerical value between -9,223,372,036,854,775,808 to 9,223,372,036,854,775,807. Its default value is 0. |
| Short | This data type allows you to store a numerical value between -32,768 to 32,767. By default, it is set to 0. |
| String | This data stores alphanumeric values. Empty string variables will give null values by default. |

## Creation of Variables

We need to write a declaration statement in order to create a variable for a certain data type. Any name can be assigned to a variable which we will also use when calling the stored data in the future block of codes.

Let us create a variable using the integer data type and let us name this variable as "MyAge". Our declaration syntax should be:

int MyAge;

By default, the program will assign the zero value to the MyAge variable since we did not specify an integer value. In order to assign a specific value inside the MyAge variable, we will use the equal ( = ) operator sign. Let us say that we want to assign the value 20 to the variable. Our syntax will now look like this:

int MyAge = 20;

The same goes with other data types. Here is the list of examples for each data type when declaring them in our IDE: (Assigned variable name is "TestVar")

| Syntax Formula | Example |
| --- | --- |
| Boolean [variable name] = [true or false]; | Boolean TestVar = true; |
| byte [variable name] = [numerical value]; | byte TestVar = 0; |
|  |  |

| | |
|---|---|
| char [variable name] = [single character]; | char TestVar = L; |
| double [variable name] = [numerical value].[numerical value]d; | double TestVar = 31.13d; |
| float [variable name] = [numerical value].[numerical value]f; | float TestVar = 1.3f |
| int [variable name] = [numerical value] | int TestVar = 5 |
| long [variable name] = [numerical value] | long TestVar = 0L |
| short [variable name] = [numerical value] | short TestVar = 32000 |
| String [variable name] = [alphanumeric value] | String TestVar = null |

## Data Type and User Input

```
Editeur - [Java syntax text editor.jav]
File Edit Search Macro Tools Window Help

public class CreateObjectDemo {

    public static void main(String[] args) {

        // create a point object and two rectangle objects
        Point origin_one = new Point(23, 94);
        Rectangle rect_one = new Rectangle(origin_one, 100, 200);
        Rectangle rect_two = new Rectangle(50, 100);

        // display rect_one's width, height, and area
        System.out.println("Width of rect_one: " + rect_one.width);
        System.out.println("Height of rect_one: " + rect_one.height);
        System.out.println("Area of rect_one: " + rect_one.area());

        // set rect_two's position
        rect_two.origin = origin_one;

        // display rect_two's position
        System.out.println("X Position of rect_two: " + rect_two.origin.x);
        System.out.println("Y Position of rect_two: " + rect_two.origin.y);

        // move rect_two and display its new position
        rect_two.move(40, 72);
```

We already know that we can collect input data with the use of Scanner. Now, we will use the Scanner class to collect other data types. In this example, we will ask for a numerical value from the user and then ask the program to save the value as an integer value.

Our first step is to tell the Scanner class that the value it needs to store is a value of the integer data type. We can do it by changing the nextLine() parameter into nextInt() parameter in our output.

Let us create a program where we will ask the year of birth of the user. Afterward, let us return that value to the user as an integer data type. Our IDE should be like this:

| | |
|---|---|
| 1 | |
| 2 | package javatestproject; |
| 3 | import java.util.Scanner; |
| 4 | |

```
5    class JavaTestProject {
6
7       public static void main(String[] args) {
8
9           System.out.print("What year were you born?");
10          Scanner BirthYear = new Scanner(System.in);
11                          System.out.println("You were born in "+BirthYear.nextInt());
12
13      }
14
15   }
16
```

## Storing User Input into Another Variable

When calling the Scanner class to collect data from user input, we create a variable as well. In our Java program with regards to the birth year, for instance, we have created a new variable named as "BirthYear" and we store the collected data in it. The problem here is that we also write the parameter right after the variable name when executing the Scanner class data in our program's output. This is too much hassle when you need to call the collected data in multiple statements. Hence, we can store the entire syntax into another variable to simplify our codes. Do this by declaring a variable while assigning the Scanner syntax as the variable's value.

Let us create another program where we calculate the user's age five years from now. This time, we will also store the Scanner call statement into another variable. This is how it should look like:

```
1
2   package javatestproject;
3   import java.util.Scanner;
4
5   class JavaTestProject {
6
7       public static void main(String[] args) {
8
9           System.out.print("How old are you?");
10          Scanner UserAge = new Scanner(System.in);
11          Age = UserAge.nextInt();
12          System.out.println("You will be "+(Age+5)+" five years from now");
13
14      }
15
16  }
17
```

From our example program, we stored the Scanner call statement into a new variable which we named as "Age". On the next line, we called the user collected integer and wrote a simple arithmetic operation in order to provide how old our users will be after five years from now. Notice that the mathematical expression is enclosed in parentheses. Without the parentheses, the program will connect the stored value in the Age variable with the numerical value 5 instead of adding the two values together.

In addition to integers, you may also use the following parameters to look for other types of data using the Scanner class call statement:

- nextByte() for Byte data type values
- nextShort() for Short data type values
- nextLong() for Long data type values
- nextFloat() for Float data type values
- nextDouble() for Double data type values

# Chapter 3 – Operators

Since the first chapter, we have been using different operators and the most common operator we have used is the assignment operator. It is defined by the equal ( = ) sign and it signals the program to designate a certain value to a variable.

Aside from the assignment operator, there are five other operators that you use to manipulate data in the Java programming language. They are as follows and their role as mathematical operators:

- Operator for Addition which is represented by the plus sign ( + ). It commands the program to give you the sum of multiple values.

- Operator for Subtraction which is represented by the hyphen sign ( - ). It commands the program to give you the difference of multiple values.

- Operator for Multiplication which is represented by the asterisk sign ( * ). It commands the program to give you the product of multiples values.

- Operator for Division which is represented by the slash sign ( / ). It commands the program to give you the quotient of multiple values.

- Operator for Remainder which is represented by the percentage sign ( % ). It commands the program to divide two numerical values and gives you the remainder of the numerical values.

## Summary in a Single Program

The best way to familiarize with the operators is by creating a Java program where we can use at least one of the mathematical operators. We will also integrate what we have learned so far. In other words, we will also use the Scanner class to collect data from the user and then execute an arithmetic expression to provide efficient output back to the user.

In this program, we will ask the user to input two numerical values and we will let the program to provide the sum of the numerical values. Let us begin writing our Java program:

```
1
2    package javatestproject;
3    import java.util.Scanner;
4
5    class JavaTestProject {
6
7        public static void main(String[] args) {
8
9            int Value1, Value2, Total;
10           System.out.print("Please enter the first value");
11           Scanner FirstValue = next Scanner(System.in);
12           Value1 = FirstValue.nextInt();
13           System.out.println("Please enter the second value");
14           Scanner SecondValue = next Scanner(System.in);
15           Value2 = SecondValue.nextInt();
16           Total = Value1 + Value2;
17           System.out.println("The total of the two values is "+Total);
18
19       }
20
21   }
22
```

In this program, you will notice that we declared three variables as containers for integer values. This was written on line 3 where we used the following syntax:

> int Value1, Value2, Total;

Because of this, we no longer need to enclose the two variables inside the parentheses as we call them for a mathematical expression on line 16.

We also allow users to enter two numerical values by using two Scanner call statements. Then, we stored these values into two separate variables that added to one another inside another variable. This way, we were able to call the sum of the two values for our final output stream.

# Conclusion

The Java programming language gives you unlimited tools to achieve different operations. However, it is extremely important to get to know the IDE you chose to use. Beginners should find which IDE they feel most comfortable working on. Although all IDEs use a single programming language, each has its own unique interface that might change your programming pace.

We also learned how to start a new project and the essential elements a project needs such as the main class and the main method. Packages were also introduced and although the package declaration is optional, we have learned that declaring packages in our Java program will help us efficiently organize multiple classes for bigger projects.

The main class and the main method may also contain access modifiers that will allow the program in defining whether it should allow other classes in accessing the statement or not. It is a general practice to use "Public" as the main class's access modifier.

Allowing the program to return a message as its output has also been discussed in the first chapter of this book. You may use the keyword "print" as the simplest form of the parameter or you may rather sign the Java program to show the string in another line. This can be done with the use of the keyword "println".

We also discussed how dull it is to have a one-way-only program where users were not capable of sending an input. The Java programming language has an answer to this by allowing you to use its built-in Scanner class which can be found in the built-in local package known as java.util. The Scanner class allows us to collect data entered by the user from their keyboard. This opened new operations such as returning the user input as the program output.

The next chapter introduces variables and how do they bring out the full potential of the Java programming language. We discussed the different data types that we can store inside variables and how to call these data for our program's output. It is also important to remember that we can also store variables into another variable for efficient programming later on.

Lastly, our last chapter discusses the different operators aside from the assignment operator represented by the equal sign ( = ). In the end, we combined everything we have learned to create a Java program which can solve a mathematical equation based on user's input.

At this point, I am confident that you already grasped the idea behind the Java programming language as an object-oriented language. I personally congratulate you for finishing this book and I am quite sure that you are now ready to move forward to a more advanced method in Java programming.

Do not forget that you have only took a glimpse on what the Java language is capable of. This book has a continuation where you will learn more advanced techniques such as flow control, a deeper understanding of access modifiers, a closer look to objects and classes, constructors, serials, and inheritance.

Consider this book as your big leap into the world of Java programming language. Always remember that this programming language is fun to learn, yet filled with tons of challenges ahead. Nevertheless, mastering the language will open new opportunities to you such as being able to develop your own mobile application. You may even end up developing a new stand-alone game that may stand the test of time like Minecraft.

# C++

## The Ultimate Guide to Learn C Programming (c plus plus, C++ for beginners, programming computer, how to program)

PETER HOFFMAN

# CONTENTS

Introduction ................................................................................. 43

Chapter 1 – WRITING YOUR FIRST C++ PROGRAM ...................... 45

Chapter 2 – EXPANDING YOUR PROGRAM: VARIABLES ............ 50

Chapter 3 – OPERATORS ........................................................... 57

Chapter 4 – CONDITIONS and FUNCTIONS ............................... 60

Chapter 5 – LOOPS ................................................................... 66

Chapter 6 – ARRAYS ................................................................. 74

Chapter 7 – POINTERS .............................................................. 80

Chapter 8 – DYNAMIC MEMORY ............................................... 84

Chapter 9 – CLASSES AND OBJECTS ........................................ 89

Conclusion ................................................................................ 95

# Introduction

A C++ program is simply a text file which contains a sequence of commands using specific C++ text. We can call this text file the *source file*, and it will normally end in a .CPP extension (just as an audio file ends in .MP3 or a PDF document ends in .PDF).

The purpose of writing a C++ program is to put together a specific sequence of C++ commands which will be converted by the computer into *machine-language* which will perform whatever task that we want done. This conversion process is called *compiling* which is performed by the compiler. In addition, the code that we write must be written to include both setup and tear down instructions in a process which is called *linking*. Together, the process of compiling and linking is called *building*, which builds our code into a machine-executable .EXE file of which the computer can run as a program that it understands.

In order to write a program in C++, you will need only two things – an editor to write the code and build the .CPP file, and a compiler which will convert the code (the source file) into a machine-executable file which will do the tasks that we have programmed it to do (such as open a file, play a video, print text on the screen, and so on).

This book will use examples for you to follow along with using Microsoft Visual Studio – the perfect companion to both the beginner and more advanced C++ programmer.

# Chapter 1 – WRITING YOUR FIRST C++ PROGRAM

Before we write our first program in C++, you are going to need to download the software called Microsoft Visual Studio, which allows you to write, compile and execute commands and programs written in C++ code language. It is also an excellent program which can be used to debug and edit your program should it need changes or if something goes wrong.

As your first program, we are going to explore the basic "Hello World" program which is a must for everybody who is wanting to learn how to write and code C++. The Hello World program will demonstrate the absolute bare essentials that nearly every program written in C++ will share. It does not matter how simple, or how complex a program is, it will still require the following few lines of very basic code.

The first step you need to do, is open Microsoft Visual Studio and create a new project. We will call this project "Hello World" or "First Program" if you prefer. Don't worry if you do not understand the following lines of text just yet, as we will have a closer look at them in a few moments and explore them deeper in the chapters which are to follow.

Firstly, you should write the following lines in Microsoft Visual Studio:

#include <iostream>

using namespace std;

These first lines are basically what we call a "pre-processor directive", or in other words, letting the code to follow know that we are going to start a program. The letters "i" and "o" in the word <iostream> stand for both input and output, which is required in every C++ program that is written. The final line, "using namespace std", means that we are telling the computer that we will be using the standard library (a library in this case meaning the type of commands we are using and writing). In other words, we are tell the compiler (Microsoft Visual Studio), to basically "include" the library called "iostream". This library is what holds all of our functions that we use, but more on that in a minute.

The next lines that you should write are:

int main () {

These two lines tell the program to start. It basically means that we are showing the "start" of the program to the compiler. The { is very important here, and is known as a "Scope Bracket". It is important, and will become very important as we continue to write this program and any others that exist, because every single line of code that you write must end in a ; or { / }.

The next lines that you should write are:

cout <<"Hello World"<<endl;

This line is very simple. The word "cout" is a function which we are using from the iostream library, and simply works as a print statement. In other words, by using the cout function, you are telling the program that you would like to print (display on screen) the words "Hello World". Feel free to change these words to whatever you want. The line ends in "endl;", which is where we are basically telling the program that we are going to be writing another line after this one. It is basically the same as pressing the Enter key on your computer keyboard to write another line.

The second last line of code should be:

return 0;

This is where we are telling our compiler to "return" back to the command prompt (which we will discuss later in more depth) if the program ran smoothly without any problems.

Finally, the last line of code should be written as:

} //end main

The } is basically the closing bracket. As we discussed above, all commands should end in a Scope Bracket which tells the compiler that the command is finished (for that line, or for the whole program). The last part, "//end main" is a function that we call a "comment". These are what we use to keep track of what we are using and writing in the program. In the case where we are writing a very long program, this final statement is basically the closing bracket to the function that we started with, int main(), which will keep the whole program together as one entity.

Now, our program should look like this:

```cpp
#include <iostream>

using namespace std;

int main () {

cout <<"Hello World"<<endl;


return 0;

} //end main
```

Now when we save our program and then run it, it will display the words "Hello World". Try it again, except this time I want you to write something else and see what happens.

In the next chapter we will start to look at some of the major features and functions of a program.

## Chapter 2 – EXPANDING YOUR PROGRAM: VARIABLES

The most important part of C++ code is a *variable*. You can look at these as if they were small storage boxes where we can store things for later use by the program – especially numbers. The term and use of these variables comes from mathematics. You should be familiar with something like this:

x = 1

This means that x is equal to the number 1. The statement is also saying that wherever this letter x appears, it will be worth the number 1. In essence, the letter x is *holding* or *storing* the number 1 until it needs to be used in a statement or algorithm. Variables work in exactly the same way in C++, except for the fact that we will always include a scope bracket at the end to imply and tell the program that it is the end of a function. For example:

x = 1;

From this point on, wherever we use the letter x in the program, it will be recognised as the number 1. But how would be declare variables? Generally we are not going to just use one variable when it would be much easier to just write the number. Lets look at an example from algebra that you might see in a more advanced level of mathematics:

$(x + 2) = y / 2$

$x + 4 = y$

solve for x and y

If you are any good at algebra you may have already solved the problem. However, simply inputting this information into C++ is not enough for the program to understand what this all means. It will read it only as gibberish and it will make no sense at all to the computer. This is why we must *declare* our variables to the program before we can use it, so that the computer knows exactly what we are trying to do. So we would declare our variables by:

int x;

x = 10;

int y;

y = 5;

As we have seen in the previous chapter, the statement "int" basically means "start". Here, we are saying that x equals 10 and y equals 5. If we were then to write the code again, the program would be able to solve the problem. We can use these variables anywhere that we like in the program, but we must always declare to the system what they mean before we can use them. And it is not only numbers that we can store in the variables. We could also use the following:

x = 1

x = 2.3

x = "this is a sentence"


To give you a further example of how variables work, consider the following. I tell you to remember the number 5 and the number 2. You have just stored two values in your mind. I then tell you to add the number 1 to the first number which I said. Now, in your memory, you should have the number 6 (5+1) and the number 2. I then ask you to subtract the first number from the second and then we get the answer of 4.


This basically explains how C++ will store variables and then use them to solve problems or statements. When actually writing it in code, it would look like this:


a = 5;

b = 2;

a = a + 1;

result = a – b;


This is a very simple statement which you could have probably solved faster than writing it, but just remember that your computer has the potential to calculate billions of these problems every second, so you should be able to see the bigger picture of what each of these variables can do for you and larger programs.

But what if we wanted to write a small program which we could practice declaring and using variables? There are two ways which we could write this. We can write in a single statement such as:

int a, b, c;

Which means that a,b,c are variables with the type "int", which means the same as:

int a;

int b;

int c;

It depends on whether we want to type more than one line, but both represent the same statement. To use both in an example, and to see our mathematical problem we just solved using variables, lets have a look at what it would look like in a small program on the next page.

// operating with variables

```cpp
#include <iostream>
using namespace std;

int main ()
{
// declaring variables:
int a, b;
int result;


// process:
a = 5;
b = 2;
a = a + 1;
result = a – b;


// print out the result:
cout << result;


// terminate the program:
return 0;
```

}

When we run the program, we will see that it will display the answer for us. Very simple and you should be able to interpret what each of it means using what we have discussed so far with our "Hello World" program and the variables we have used. However, we do have something new here, and that is the use of "//" followed by some text. The first "// operating with variables" is simply a title for our program, and each of the following // are simply sub-headings explaining what we are doing at that point of the program. It makes it a lot easier, especially if you are dealing with a very large amount of code, to go back through the lines and find what you are looking for instead of having to try and decipher each line.

There is also an easier way to write the code for the example that we just used. This second method is called the "initialization" of a variable, and there are three ways that we can do it. To *initialize* a variable, basically means that we can declare its value at the same time that we declare the variable. The three ways that we can do this is by writing one of the following:

int x = 0;

int x (0);

int x {0};

It does not matter which method that you use because they all do the same thing. The best part however, is that you can use all three types at once if you chose to. For example:

```cpp
// initialization of variables

#include <iostream>

using namespace std;


int main ()

{

int a=5;

int b(3);

int c{2};

int result;


a = a + b;

result = a - c;

cout << result;


return 0;
```

}

As you can see, the program will do exactly the same thing that we wrote it to do before, except for the fact that now we have declared the value of the variables as we have declared the variables themselves.

# Chapter 3 – OPERATORS

Now that we are familiar in using variables, we can expand on our program and start using what are called "operators". Operators are basically symbols which tell the compiler (Microsoft Visual Studio), to use mathematics or logic to solve the problems that we have written in code. There are 6 types of operators which C++ will most commonly use. These are:

1. Arithmetic Operators.
2. Relational Operators.
3. Logical Operators.
4. Bitwise Operators.
5. Assignment Operators.
6. Misc Operators.

We will look at only the main and most important types of operators and define what each of the most commonly used symbol in each mean and what they do.

## ARITHMETIC OPERATORS

Let's imagine that we have declared that A equals 10 and B equals 20. We could use the following arithmetic operators to do the following with those variables:

| OPERATOR | WHAT IT DOES | EXAMPLE OF USE |
|---|---|---|
| + | Adds together | A + B = 30 |
| - | Subtracts #2 from #1 | A – B = -10 |

| | | |
|---|---|---|
| * | Multiplies the numbers | A * B = 200 |
| / | Divides the numbers | B / A = 2 |
| % | Gives a percentage | B % A = 0 |
| ++ | Increases integer by one | A++ = 11 |
| -- | Decreases integer by one | A-- = 9 |

## RELATIONAL OPERATORS

Using the same variables as above, we could also use:

| OPERATOR | WHAT IT DOES | EXAMPLE OS USE |
|---|---|---|
| != | If equal, statement is true | (A != B) is true |
| > | Is left # higher than right? | (A > B) is not true |
| < | Is left # lower | (A < B) is true |

|  |  | than right? |  |
|---|---|---|---|
|  | >= | Is left # greater or = to right? | (A >= B) is not true |
|  | <= | Is left # lower or = to right? | (A <= B) is true |
|  | (A == B) | If not equal, not true | (A == B) is not true |

These are the main two types of operators which are used 90% of the time, and it is important that you get to know what each of them mean before you start looking into the more advanced and complex types of operators.

# Chapter 4 – CONDITIONS and FUNCTIONS

Conditionals are what gives the program choice to choose what it would like to use. This doesn't mean that the program can think for itself and make decisions like a human being, but rather it makes a choice based only on the options of which you write for it to choose from. The makes programs written in C++ a huge amount of versatility, and stops you from designing a program which only keeps doing the same thing over and over again. Imagine if our "Hello World" program did nothing but keep saying "Hello World"? Or if our variable program did nothing but give us the answer "4" all of the time? These choices are smaller programs within our larger program which we call "functions". What that means is that in order for a program to complete its task, it must run through a series of smaller tasks and choices (conditions) in order to output a result. To put simply, everything that we tell the program to do as a whole is a function.

To look at Conditions, to give our program choice we will use a syntax like the following:

if (and then we include our condition here)

{

insert our program here for the program to work out

}

This is basically saying, "If this happens, then do this". You should be familiar with the basic terms of IF, THEN, AND, OR which is basic mathematics and problem solving skills. However, it is within the brackets where we put our condition, where the program can choose whether or whether not it is going to execute that condition. For example, if we have the three variables of x,y and z, we will first need to declare their values to the program. However, to start including some operators into the equation, lets say that the variable "x" is greater than the variable "y".

Our code would look like this:

```
int x,y,z;

cin>>x;
cin>>y;

if(x>y)
{
    z = x+y;
    cout<<z;
{
```

Once we declare the values of the variables of x and y, the program will determine if "x" is indeed greater than "y". If the program finds this to be true, then the undeclared variable "z" will be declared as the sum of x and y and then give us the answer. However, if the program finds that x is not greater than our new variable of z, nothing will happen.

This is where we give our program choice of what it can do when faced with different scenarios. In this example, we must now tell the program what to do if x does not equal a number greater than the variable z.

Look at the following:

int x,y,z;

cin>>x;

cin>>y;

if(x>y)

{

z = x+y;

cout<<z;

{

else if (x<y)

{

z=x-y;

cout<<z;

}

Now we have told the problem that if x does not equal more than z, to go to the next option which tells us to give the answer of the number of the result of x minus y. You can add as many of these conditions as you like which will allow you to create a very versatile program which can be applied in many ways. But, using our example, what if variable x is not more or less than variable y? Nothing would happen. We would still want to program to do *something*, so we could include:

```
//Full Working Program

#include <iostream>

using namespace std;

int main()

{

int x,y,z;


cin>>x;

cin>>y;


if(x>y)

    {

    z=x+y;

    cout<<z;

    }
```

```cpp
else if (x<y)

    {

    z=x-y;

    cout<<z;

    }

else

    {

    cout<<"x is not greater or less than y so they must be equal"<<end1;

    }

return 0;

}
```

Now we have created a program in which if x is not greater or less than the variable y, then the "if" and "else if" conditions will not be executed. The program now has the choice to execute the third condition of "else", since the first two conditions can not be met. We can also use many different types of operators in these strings of conditions. Refer to the previous chapter on Operators to see those.

# Chapter 5 – LOOPS

We have looked at basic programming, variables, operators and conditions. Now we must have a look at loops. Loops are a very important part of programming, as they will give our program the ability to keep running the same lines of code as many times as you tell it to. You may think that doing the same task over and over again is pointless and useless, but the meaning behind loops will become very clear to you as we look at the following chapter.

Let us look at the following lines of code from a program.

#include <iostream>

using namespace std;

```cpp
int main()
{
cout<<"1"<<endl;
cout<<"2"<<endl;
cout<<"3"<<endl;
cout<<"4"<<endl;
cout<<"5"<<endl;
cout<<"6"<<endl;
cout<<"7"<<endl;
cout<<"8"<<endl;
cout<<"9"<<endl;
cout<<"10"<<endl;

return 0;
}
```

As you can probably see, based on what we have learnt so far, is that this simple program will display for us the numbers from 1 through to 10. You can probably guess that if you were to have to include a lot of these lines in your program, the task would become very boring and monotonous. Can you imagine if you had top design a program which would display the numbers from 1 through to 1000? Luckily though, there is a solution to make our job a whole lot easier. We can create a syntax which will turn the above example into one simple line of code. This syntax will be a loop, and will look like this:

for (initialization; condition; increment/decrement)

This is basically saying, "for the following, do this". The initialization is the start of the program (as you should know by now), the condition is whatever choice we give to the program to execute, and lastly, the increment and decrement will represent by how much we would like to add or subtract from our starting number as the program and loop continues to run.

The following is the same example as above, except this time we have included a loop.

```
#include <ostream>

using namespace std;

int main()
{

    int x;
```

```
for (x=0; x<=10; x++)

    {

        cout<<x;

    }


return 0;

}
```

This program will execute exactly the same as the example above to print numbers, except in this one, we did not need to specify every single number between 1 and 10 for the program to display.

What makes this new set of code more effective, is that we can see that we have used "for (x=0; x<=10; x++)" which is our (initialization; condition; increment/decrement) statement that we looked at above. Here, we have declared x to equal 0 when the program starts, and the program will end when x equals less than or equal to 10. The increment/decrement of "x++" means that the value of x will increase by the number 1 at the end of every loop.

To explain it further, the condition for the loop to run is that the program asks itself, does x equal 10 or less than 10? If yes, 1 is added to x. The program will then ask itself, x is now worth 1, is that equal to or less than 10? Yes, so I will add another 1. Eventually the program will say to itself, "x equals 10, is this equal to or less than 10? Yes it is. I will not make another loop and will now end the program". This is basically how loops work. The program will have a defined amount of conditions that will keep being checked upon, and if those conditions are passed or met, there will be another loop or repeat of the sequence.

What we have looked at is called a "for" loop. There are two other types of loops that we will also look at. The first new loop that we will look at is called a "while" loop. A while loop will allow your program to be more versatile in some circumstances, and the basic syntax looks something like this:

while (condition)

{

statements

}

This basically says, "while these conditions are true or false, I will run the loops which have been set out for me in the statements that have been made".

What would our "for" loop program look like with "while" conditions instead?

//Full Working Program

```cpp
#include <iostream>

using namespace std;

int main()
{

int x=0;

while (x<=10)
{
    cout <<x<<endl;
    x++;
}

return 0;
}
```

Our program will perform the same task as before, except this time it will use a "while" instead of "for" conditions for its loops. In this example though, we have declared the value of x to be 0 at the start of the program instead of writing it in our statement section. This is because a while loop is only focused on the conditions that have been set for it, and the incrementing of the number between loops. We could also simplify the while loop by writing the following:

```
while (x<=10)

{

cout<<x++<<endl;

}
```

This is saying that while x equals less than or is equal to the number 10, it will add one number. The only difference here is that x has been declared at the start of the program.

The third type of loop that we can look at is called a "do while" loop. It will perform the same task, but the condition is placed at the end of the loop, so the loop will not be checked to see if it should be performed until it runs for the very first time. For example:

```
//Full Working Program

#include <iostream>

using namespace std;


int main()

{
```

```
int x=0;

do

{

    cout<<x<<endl;

    x++;

}while(x<=10);


return 0;

}
```

This is saying that 1 will be added to x while x equals less than or equal to 10. Did you notice the scope bracket (semicolon) at the end of the while loop? This means that we are closing this line off as a function for the program to execute.

# Chapter 6 – ARRAYS

Arrays are very similar to variables in that they allow us to store information for later use. Arrays allow us to store a collection of data under a single name, which means that we have a greater level of control in organizing our data which becomes very useful especially dealing with large programs with thousands of lines of code.

To give you an example of how a set of data could be stored in an array, imagine that you wanted to design a program which would display the age of 5 people. You would need to declare the name of the variable and the value of the variable, which would look something like this:

int age1=15;

int age2=24;

int age3=54;

int age4=33;

int age5=120;

We are now faced with the same problem as before in the previous chapter where we wanted to print the numbers from 1 through to 10. Imagine if you had to design a program to display a list of the ages of 500 people? Luckily though, just as we saw in the previous chapter, there is a way to display this information without having to manually type in every single line of code.

The syntax for an array looks like this:

*datatype* name [index];

This means that we want to list a series of stored values much like a variable. In order to create an array which will store something similar, and in this example the ages of our 5 people, we would write:

int ages [5];

This means that we have a list of 5 things which the program knows only to be something called "ages" at this point. Just like declaring a value for the letter "x", we must also declare values for this new variable we have called "ages". We can do this by:

int ages [5];

ages[0]=15;

ages[1]=24;

ages[2]=54;

ages[3]=33;

ages[4]=120;

Although I mentioned that there was a much easier way to write this index, the above looks exactly the same. The reason for this is that I want to show you how we declare this new datatype of ages. When we start the program, we have told it that when it starts there is an array of [5] items. The reason that our list of declarations start at 0, is because this is how computers read a list of numbers. A 0 will always mean first.

However, if we want to write the above in a much more simpler way, we can write the following:

int ages [5];

ages = {15,24,54,33,120};

This is telling the program the same as above, in that 0 equals 15, 1 equals 24, and so on. What does this look like if we were to write it out completely in code?

# C++

```cpp
#include <iostream>
using namespace std;

int main()
{
    int ages [5];
    ages={15,24,54,33,120};
    cout<<ages<<end1;


    return 0;
}
```

Can you spot the error in the above code? Most would think that we could write this program the same way as our previous examples, in that we define a set of variables, and then have the program list those when it is executed. However, this is not the case when dealing with arrays. The problem in this code is that we have ended the program with "cout<<ages<<end1;" after we have defined the array. The problem with this is that C++ cannot print a whole array, since you must tell your compiler (Microsoft Visual Studio) what array to display. For example, if we wanted to display the age 120, we would need to tell it to display age [4]. Then we would need to list every variable as a list like as follows:

```
int main()
{
int ages [5];
ages={15,24,54,33,120};
cout<<ages[0]<<endl;
cout<<ages[1]<<endl;
cout<<ages[2]<<endl;
cout<<ages[3]<<endl;
cout<<ages[4]<<endl;

return 0;
}
```

But I thought that you said that there was a away to store a whole set of integers into an array? There is, and the way that we do that is to declare a variable, say "k" and declare to the program that it is storing integers to be used.

For example:

```
//Full working program
#include <iostream>
```

C++

using namespace std;

int main ()

{

int k;

int ages [5];

ages = {15,24,54,33,120};

    for (k=0; k<5; k++)

    {

        cout<<ages[k]<<end1;

    }

return 0:

}

Notice how we have defined that "k" is a variable and was declared to hold the entire string above? Basically, "k" was declared as 0 which will execute the array as long as "k" remains less than 5 (not equal to in this case if you notice the different operator used). The program is basically saying when it will display the ages until the datatype "age" reaches 5, which is then when the program will stop displaying the ages.

# Chapter 7 – POINTERS

Pointers are something that we have not yet looked at in our previous examples. The reason that they have been left for now is for the fact that it can be sometimes very difficult to understand exactly what they are for. This does not mean that you do not need to learn about how they work and how to use them because they are still a very important part of the structure of more complex programs.

Let's start looking at them by looking at a syntax of how we would declare a pointer:

double *p;

In this statement, we are saying that we have a pointer called "p" which can have an an address or location of floating point values. The word "double" is our new datatype, and the reason that there is an * before p is because this is how we declare a pointer.

In code, we would declare an address which a pointer points to by writing the following:

p=&x;

The letter "p" will now store the location of "x" (which is a variable in this case). In simpler terms, p is pointing to x. We use "&" before the variable in order to assign it an address (but not a value). P will point to x, but x will not have a value. To give x a value, we will need to make changes to p, and can write code to do so like this:

*p=25.3;

This now means that x equals 25.3 by making p a pointer, and by pointing p to x. Since p holds x, x is now valued at 25.3 (whatever value p is). If we were to write a complete program which uses a pointer and assigns it a value, we could write the following:

//Full working program

#include <iostream>

#include <stdlib.h>

```
using namespace std;


int main()

{

double x;

double *p;

p=&x;

*p=25.3

cout<<x<<end1;


system ("pause");

return 0;

}
```

This program points p at x and tells the program that x is valued at 25.3. However, this does not really seem that pointers are very useful. We could simply just use variables and define them (and perhaps use some conditional loops) in order to make our program work. What they are useful for though, is that they give you the ability to manipulate the information and code simply by pointing a variable at another variable. This is where they do become useful.


Let's say for example that you have written several thousand lines of code. Now, imagine if all of a sudden you need to change the value of 50 of those variables within the code? It would take a very long time to find all of those values within the lines of code, and would perhaps even mean that you would have to write the code all over. If you accidentally change something in the process when you should not have, the program may be damaged.

Luckily, this is where pointers actually have a useful and important function. We can use pointers to change the value of variables which already exist in the program, or we can point and declare the variables at the start of writing the program. Pointers allow us to manipulate a function, whereas this is impossible with a normal variable.

This is all that you need to know about pointers at a beginner level.

# Chapter 8 – DYNAMIC MEMORY

Everything, even lines of code, need storage space to be stored and to be executed. For example, when you are copying a song or a photo from your computer to your phone or vice versa, there will be a program which will ask the phone or the computer if there is enough space before the file is moved. This is because if there is not enough storage space to store the new file, it is pointless even trying to move it. The program will determine before the process is executed if there is enough storage space for the operation to be executed.

However, there are cases when the program will not know how much memory or storage is needed until the program is executed, and this amount of memory can constantly change. This is why when you write any C++ program, you need to tell it to allocate a certain amount of memory or storage to accommodate for this (and this can sometimes be unknown). This second type of memory is what we will be focusing on here, since it is easy to determine memory needs of storage before a task is executed.

Every time the program is executed, the allocated memory needs can be referred to as the "new" operator. But, as soon as the memory allocation is no longer needed, we refer to "delete" operator. The syntax for this looks like:

new datatype;

This basically means that we are defining a new operator, followed by whatever it is that we define afterwards. As an example, we can write some code where we will use a pointer to allocate memory as soon as the program is executed:

double* pvalue = NULL;

pvalue = new double;

However, these lines of code will not tell us if memory has been successfully allocated it is only a command to allocate. In that case, to have the program tell us that memory has been successfully allocated, we would write:

double* pvalue = NULL;

if( !(pvalue = new double ))

{

cout << "Error: out of memory." <<end1;

exit(1);

}

The program will now say to us, "Error: out of memory", should there not be enough available memory to allocate to the program. Once we reach a point in our program where the expected memory requirements are deemed to not require allocated memory, we delete operator. For example:

delete pvalue;

Simple as that. Now, let's have a look at what a program would like like that first allocated memory using a new operator, executed a pointer to the allocation of that memory, and then de-allocated that memory with delete operator:

#include <iostream>

using namespace std;

int main ()

{

double* pvalue = NULL;

pvalue = new double;

*pvalue = 25677.99;

cout << "value of pvalue : " <<*pvalue << endl;

delete pvalue;

return 0;

}

Running this code would show us:

Value of pvalue: 25678

It is quite simple to input new and delete operators, and definitely much easier than the other things that we have learnt in this book so far. But let's have a further look at other ways we can allocate memory to our program. For example, let's assume that we want to make a program that will display to us a line of 100 characters. Based on the examples above, the code would look like this:

char* pvalue = NULL;

pvalue = new char[100];

This is saying that we declare our pointer to be NULL, and that we have executed a new request for memory to be allocated for our 100 characters. Once we do no longer need this memory to be allocated, we insert a delete operator:

delete [] pvalue;

Now we have told our program to delete our array which we have pointed to in the previous lines of code. But, what if we wanted to allocate memory for multiple arrays at the same time? We do not need to start a new operator for each one. Instead, we can write the following:

double ** pvalue = NULL;

pvalue = new double [4] [5];

Here, we first declare that our pointer is valued as NULL, and then we tell our program that we want to allocated memory for an array with the dimensions of 4x5. Deleting is the same as before:

delete [] pvalue;

# Chapter 9 – CLASSES AND OBJECTS

Now that we have learnt the basics of programming in C++, it is time to have a look at something slightly more advanced. Classes are part of what we call "object orientated" programming. This allows us to use programs in everyday life by using objects. Objects can be defined as anything that we use in everyday life – such as cars, computers, mobile phones, and so on. As we should have learnt by now, a program is made up of very small programs within it, just the same as a car is made up of very small parts which work together in order for the car to operate. Some of these components can be changed, like turning the lights on and off in the car, whereas others cannot be changed, such as the central computer of the car. These can be differentiated by what we call "private" and "public" variables, or in other words, what you are allowed to change, and what you are not (but we will look at that in a moment).

Classes exist on a slightly more advanced level when learning C++ because they are more involved. Learning them however is a great reward, as it will allow us to create and develop programs which we can apply for use in everyday life. They also allow use to reuse code and apply it to different scenarios. For example, every player in a football team with have a name, age and an income, but each of these will be different for every player. In this case, the code would create the attribute of class which is simply the name, age and income for each player.

To write this as an example, our code for our player information would look like this:

Class Player

{

string name;

string age;

double income;

};

Our player Class now has structure. From now on, every time that we declare a value to each of the variables of the individual player, we will be creating a new record, or a new object that exists within Class Player. Every object will consist of a name, age and income. In order to declare the variables of each Player type, we use the same method as we would declaring any other variable:

Player player1;

Player player2;

Player player 3;

Once we have created a Class as seen in the first example, we can use it like any other example of data type that we have previously used. However, we want to also be able to limit what the program is able to access and manipulate within these class structures. Some objects or classes must remain the same at all times (private class), where some can be changed (public). For example, the class which contains only name, age and income must remain private, as this is a constant. On the other hand, the variables that we declare for each individual is public, because that information is not constant.

To see this in code, we would write the following:

Class Player

{

private:

string name;

string age;

double income;

public:

Player();

};

Executing the above lines of code would prohibit the program from being able to make changes to the private class of name, age and income, but would allow it to make changes to the public class of individual player information. We use ":" after declaring private and public, because they tell the program that everything afterwards is considered that class type.

When we actually input and manipulate public data, we use what are called "constructors" and "destructors". Every time that we declare a class object, a constructor is executed to perform whatever it has been programmed to do. To initiate a constructor, we would write the following code:

Player();

Player(string,int,double);

This means that we have told the constructor to use our three parameters of name, age and income. We also have a Player class with no value on the top line, which gives the constructor the option to assign or not to assign parameters to the class type. But, we still need to tell the constructors what to do, so we would write:

Player()

{

name = "";

age = 0;

income = 0;

}


Player(string n, int a, double s)

{

name = n;

age = a;

income = i;

}


The constructor that is executed first has nothing to do apart for setting everything to the value of 0. However, the constructor which is executed second must use our three parameters of name, age and income. Then, all that is left for us to do is to actually declare each Player as required.


Player player1;

Player player2("David",33,95.20);


Much more simple to understand when looking at what a constructor does when we see it in code. Basically, they create a class or an entry of information based on what we give it. It does the work for us to a degree. However, we also need to look at destructors and what they achieve within our program.

A destructor is the opposite to a constructor in that they will remove instead of create data. The syntax used to execute a destructor is ~ followed by the name of your class. For example:

~Player();

And that is the basics of classes and objects within C++ programming. There are of course a lot more which we could look at, but that scope goes beyond the realm of the beginner level.

## Conclusion

In summary, by reading this book you should now have the required basic skills required to understand the basics of the C++ programming language, and should now be able to program your own simply C++ programs with ease. Just remember that although on the outside every program looks complex and filled with many lines of code, all that it is simply is a bunch of smaller programs which make up a whole. Just focus on working on those smaller programs one at a time and you will be fine.

This book has also shown you the many different types of code that we can use when making a program in C++, and this knowledge will allow you to be able to impress your family and friends with simple programming skills, or perhaps allow you to make basic programs to use in your every day life. You should also now have the basics required to move on to more advanced topics on your way to become an advanced or professional C++ programmer.

Use this book as a guide or as a reference as much as possible on your journey through the wonderful and exciting world of computer programming, and always remember that C++ is easy to learn once you know the basics.

Thank you for reading. I hope you enjoy it. Ask you to leave your honest feedback.

Made in the USA
Middletown, DE
09 January 2019